I0499984

What is Infinity ?

Unveiling the Nature and
Operations of Indeterminate Forms

Contents

Title: What is Infinity? Unveiling the Nature and Operations of Indeterminate Forms

Have you ever pondered over the enigmatic concept of infinity? While we are familiar with basic operations involving infinity, such as $\infty + n = \infty$ or $\infty - n = \infty$, we lack a concrete understanding of why these relationships hold true. The nature of infinity remains elusive, existing as a mere abstract notion in our mathematical discourse. This book delves into the exploration of infinity, seeking to unravel its essence by examining the fundamental definition of numbers and their composition. By dissecting numbers into their constituents, we can shed light on the elusive nature of infinity and comprehend its role in three specific indeterminate forms: $\infty - \infty$, ∞/∞ and 1^{∞}.

To embark on a profound exploration of infinity, it is essential to establish a strong foundation by delving into the basics and gaining a comprehensive understanding of the true nature of numbers in our real-life experiences. In this endeavor, we shall unravel the intricate fabric of the numerical system and unravel the profound meaning that numbers hold.

Introduction

In our everyday lives, we encounter numbers and engage in various operations involving them. Yet, it is imperative to reflect upon the deeper significance that numbers possess. What, indeed, does the number 1 signify? How about the number 2? Are these numerical entities mere symbols devoid of intrinsic meaning, or do they bear profound significance?

When we consider numbers, we encounter two distinct aspects: the value they hold and their representational symbols. Each number, regardless of its symbolic representation, carries its own intrinsic value. To comprehend this concept, let us transport ourselves back to ancient times, when numbers and symbols had not yet been invented.

In those bygone days, people relied on their fingers as a method of counting. The human hand, with its five distinguishable digits, served as an intuitive tool for quantifying small quantities. However, when it came to larger quantities, a more practical approach was required. Thus, they employed stones as rudimentary boards upon which they inscribed the number 1. These stones became the bedrock of counting, facilitating all manners of calculations and numerical operations.

To illustrate the practical significance of numbers in this context, let us consider a scenario from the past. Imagine a community of shepherds who would take their sheep out for grazing in the vast expanse of the countryside. However, upon their return, they faced a challenge: ensuring that all the sheep they had taken with them had come back safely. In the absence of modern tracking methods, the shepherds resorted to a clever solution rooted in the fundamental concept of the number 1.

As a shepherd prepared to set out with their flock, they would inscribe as many instances of the number 1 as there were sheep in their care. Each 1 symbolized an individual sheep. These inscriptions on a stone or any available surface served as a tangible representation of the flock, acting as a numerical record. When the shepherds returned from grazing, they would meticulously count the number of sheep that had come back. To accomplish this, they would physically erase or remove the 1 symbols, one by one, for each sheep that safely returned.

In this way, the shepherds would effectively ascertain whether any sheep had gone missing during the grazing period. By observing the remaining 1 symbols, they could instantly identify the number of sheep that had not returned. This practical application of the number 1 as a base unit for quantification underscored its fundamental role in this particular context.

The example of the shepherds highlights the inherent value of numbers in representing and quantifying real-life entities and phenomena. While the numerical symbols may have evolved over time, the underlying concept of numbers as fundamental units of measurement and representation remains unchanged.

In our contemporary society, we continue to rely on numbers to navigate various aspects of life. From counting money and tracking time to measuring distances and calculating quantities, numbers permeate numerous aspects of our existence. Whether we realize it or not, numbers shape our perceptions, guide our decisions, and facilitate our interactions with the world.

Beyond their practical applications, numbers also possess abstract and philosophical dimensions. They enable us to explore the vast realm of mathematics and engage in profound intellectual pursuits. Concepts such as infinity, prime numbers, and irrational numbers challenge our understanding and expand the boundaries of human knowledge.

Moreover, numbers serve as a language for expressing patterns, relationships, and concepts across disciplines. In music, numbers are fundamental to understanding rhythm, harmony, and scales. In art, proportions and geometric patterns rely on mathematical principles. In science, quantitative analysis and statistical models rely on numerical data.
In conclusion, to comprehend the concept of infinity and reach a deeper understanding of numbers, we must acknowledge their significance in real-life contexts. Numbers

extend beyond mere symbols, encapsulating intrinsic values and representing tangible entities or abstract concepts. By recognizing the historical foundations of numerical systems and appreciating their applications in our daily lives, we can unlock the profound meaning and power that numbers hold. Embracing the realm of numbers equips us with a versatile toolset to navigate the complexities of the world and explore the infinite possibilities that lie before us.

The Number '1'

The Universal Core:

The concept of the base number 1 is indeed fascinating, as it serves as the foundational element upon which the entire numerical system is constructed. As we delve further into the intricacies of the numerical system, it becomes evident that each number derives its existence from the preceding number, ultimately tracing back to the base number 1.

Consider the number 2, for instance. It is inherently tied to the existence of 1. Without the presence of 1, the concept of 2 would be devoid of meaning. This can be illustrated through a relatable example: the notion of having two cars. It is impossible to possess two cars if we do not first possess one car. The number 2, therefore, relies on the base number 1 to acquire its significance.

Extending this understanding, we can examine the relationship between other numbers within the numerical system. Take the number 5, for instance. It is composed of four 1's. In turn, the number 4 is formed by three 1's, and so on. We can observe a cascading pattern where each number is built upon its predecessor, with 1 acting as the fundamental building block. Therefore, it can be concluded that every number in the universe exists due to the presence of the previous number, with 1 serving as the base.

To gain a visual understanding of the entire numerical system, one can envision a spherical structure with an infinite radius.

This ball of infinite radius is comprised of an infinite number of elementary balls, each representing the base number 1. The numerical values we encounter, such as 2, 4, or any other large number, can be visualized as combinations of these elementary balls of 1.

For instance, if we consider the number 2, we can extract two elementary balls of 1 from the numerical system. Similarly, the number 4 can be represented by four elementary balls of 1. By extending this analogy, we can comprehend that any given number, regardless of its magnitude, is ultimately constructed using the base number 1. This visualization aids in understanding the hierarchical structure of the numerical system and highlights the fundamental role played by the base number 1.

The concept of the base number 1 not only underpins the construction of the numerical system but also has profound implications across various disciplines. In mathematics, the notion of the base number is foundational to the understanding of number systems and their operations. In computer science, for instance, the binary system employs the base number 2, where all numbers are represented using only 0 and 1. This binary system relies on the fundamental concept of the base number to convey complex information through simple digits.

Moreover, the concept of the base number extends beyond the realm of mathematics and finds relevance in fields such as linguistics and anthropology. In linguistics, the study of numeral systems across different languages reveals intriguing variations in the choice of base numbers. While the decimal system, with base number 10, is prevalent in many cultures, other numeral systems may employ different base numbers. For instance, the duodecimal system uses base number 12, and the vigesimal system uses base number 20. The exploration of these diverse numeral systems highlights the significance of the base number and its influence on human cognition and cultural practices.

In conclusion, the concept of the base number 1 plays a fundamental role in the numerical system. It serves as the building block from which all other numbers are derived, embodying the essence of existence and quantification. Understanding the significance of the base number enhances our comprehension of the numerical system, enabling us to unravel the complex relationships between numbers. Whether in mathematics, linguistics, or other disciplines, the concept of the base number permeates various aspects of human knowledge and provides a framework for understanding the intricacies of the numerical world we inhabit.

You can understand the whole topic above with the given figures below:

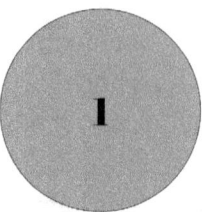

Description : This image presents a visual and imaginative depiction of the number 1, illustrating how it can be visualized.

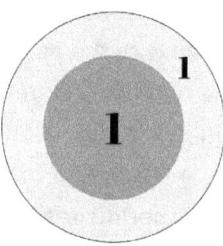

Description : This image creatively depicts the number 2 and suggests its representation as the sum of two ones (1+1).

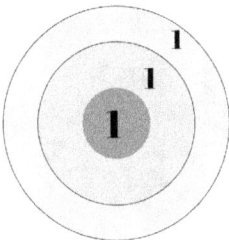

Description : This image creatively portrays the number 3 and suggests its representation as the sum of three ones (1+1+1).

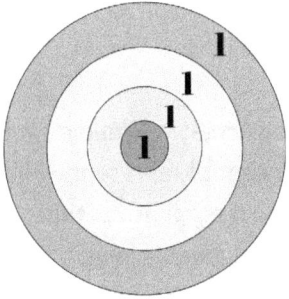

Description : This image creatively depicts the number 4 and suggests its representation as the sum of four ones (1+1+1+1).

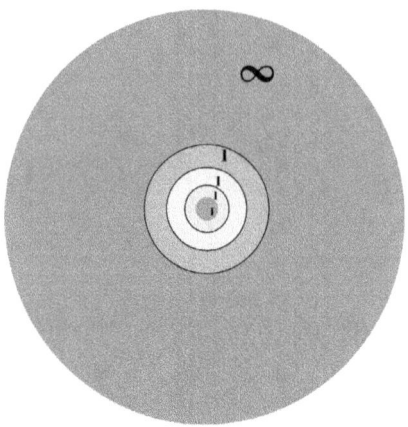

Description : This image creatively depicts Infinity and demonstrates the concept of representing infinity as the sum of 1 repeated infinitely. It emphasizes that the magnitude of infinity surpasses any imaginable large number. The purpose of this image is to aid in comprehending the overall scenario.

Smallest Number:

Introduction:

The concept of the smallest number is an interesting one, and it raises questions about relativity, perception, and the nature of numbers themselves. While we may instinctively think of smaller decimal fractions like 0.1 or 0.001 as being smaller than 1, it's important to consider that the idea of "smallness" is not absolute, but rather dependent on our human-scale perception and the context in which numbers are observed.

Relativity of Numbers:

Numbers, including their relative sizes, are not fixed entities but are rather dependent on the scale or perspective from which they are observed. Let's take the example of 0.1. In our human-scale perception, we consider it smaller than 1 because it is one-tenth of 1. However, for a being or entity that is 1/10th our size, their perception of scale would be different. To them, 0.1 would be equivalent to 1 in their perspective. This demonstrates how the perception of numbers is relative and can vary depending on the observer's context.

Our understanding of numbers is influenced by our own scale and perspective. As humans, we have a certain range of sizes and quantities that we commonly encounter, and our perception of numbers is shaped by this range. We tend to think of smaller decimal fractions as being "small" because they are lesser in value compared to whole numbers. However, this perception changes when we consider different scales and perspectives.

For example, imagine a microscopic organism that is much smaller in size compared to us. To this organism, the concept of a "small" number would be entirely different. What we consider small, such as 0.1 or 0.001, may be significant or even large for them. Similarly, if we consider beings or entities that are larger in size than us, our concept of smallness would be redefined.

The Importance of Context:

The concept of the smallest number is highly contextual. In certain situations or applications, a smaller number may be significant and considered small within that specific context. For instance, in mathematical calculations or scientific measurements, we often deal with numbers much smaller than 1, such as 0.000001 or even smaller. In these cases, the specific context determines the significance of the number and its perceived smallness.

Relativity in the Universe:

The idea of relativity extends beyond just numbers and their sizes. In the grand scheme of the universe, our perception of size and smallness becomes even more apparent. The vastness of the cosmos is awe-inspiring, and our human-scale perception may struggle to comprehend its enormity. However, from a different perspective, such as that of an omnipotent being, the scale and size of the universe may be insignificant or even manageable.

The concept of relativity and perception of numbers is intriguing when we consider the perspective of a divine entity. For such an entity, it is not necessary to create an immensely large universe. By altering the scale and size of beings within that universe, the perception of size and smallness can be manipulated. This demonstrates the subjective nature of our perception of numbers and their relative sizes.

Conclusion:

In conclusion, the question of the smallest number is not as straightforward as it may initially seem. The perception of smallness is relative, influenced by scale, perspective, and context. While we may consider numbers like 0.1 or 0.001 to be small compared to 1, this perception changes when we consider different scales or perspectives. The importance of context cannot be understated, as the significance and interpretation of smallness vary depending on the specific application or situation.

Furthermore, the concept of relativity extends beyond numbers and encompasses our understanding of the universe itself. Our human-scale perception may struggle to comprehend the vastness of the cosmos, but from a different perspective, such as that of a divine entity, the perception of size and smallness can be altered. This highlights the subjectivity of our perception of numbers and the relativity of their sizes.

In the end, the smallest number is a concept that is dependent on our scale, perspective, and context. It reminds us that numbers are not rigid entities but rather flexible and malleable, influenced by the observer and the circumstances in which they are considered.

So as an conclusion we can say any while any number like 0.000001 is smaller than 1 but for the entity which is 1/100000 it will be equal to one.

Concept of Infinity:

The concept of infinity often elicits different perceptions and interpretations among readers. Many individuals may associate infinity with an endpoint, believing that it represents an unattainable value or an ultimate destination. However, I propose a different perspective—one that suggests that all the numbers we commonly use, such as 1, 2, 3, 4, 100, or any other number or series, are inherently embedded within the concept of infinity. In essence, these numbers are born from infinity in the form of collections of 1's and eventually return to infinity. For those who find this concept challenging to grasp, I invite you to continue reading, as we will delve into a deeper understanding of infinity through its definition and mathematical calculations.

To comprehend the true nature of infinity, it is crucial to explore its definition within the realm of mathematics. Infinity can be understood as a concept that represents an unbounded, limitless, and endless quantity. It is not a specific number but rather an abstract notion that encompasses all numbers, both finite and infinite. By its very nature, infinity cannot be reached or grasped in its entirety. It transcends the confines of our finite understanding and exists as an expansive and boundless entity.

Now, let us consider the numbers we commonly use, starting from 1 and progressing indefinitely. These numbers are not separate from infinity; rather, they are intrinsic parts of it. Each number is intertwined with the concept of infinity, as they arise from collections of 1's and eventually merge back into the infinite expanse.

To illustrate this concept, let us examine the number 1. It serves as the foundation, the fundamental unit from which all other numbers are built. It is a representation of unity, singularity, and individuality. As we progress from 1 to 2, we are essentially creating a collection of two 1's. Similarly, moving from 2 to 3 involves adding another 1 to the existing collection. This pattern continues indefinitely, forming an ever-expanding series of numbers, all of which are interconnected and derived from the concept of infinity.

Consider the number 100. It is a culmination of a hundred 1's grouped together. However, even though it appears as a finite value, it is still embedded within the infinite framework. It is a finite representation of an infinite concept, bound by the constraints of our numerical system, but inherently linked to the vastness of infinity.

Mathematically, we can further explore this relationship between infinity and the numbers we use. If we attempt to count from 1 to infinity, we quickly realize that we can never reach a definitive endpoint. The sequence extends indefinitely, without any discernible conclusion. Similarly, if we try to determine the largest number, we encounter the same paradox. No matter how large a number we conceive, it can always be surpassed, as there is no ultimate limit within the infinite expanse of numbers.

The concept of infinity also extends beyond individual numbers to encompass infinite series and sets. For instance, consider the infinite sequence of numbers represented by the decimal expansion of π (pi), a mathematical constant. Although we can calculate and represent numerous decimal places of π, it is an irrational number with infinite non-repeating digits. Thus, it exemplifies how infinity manifests in the numerical realm.

Additionally, infinity plays a significant role in various mathematical disciplines, such as calculus and set theory. In calculus, concepts like limits and infinite sums are fundamental to the understanding of functions and their behavior. In set theory, infinite sets, such as the set of natural numbers or the set of real numbers, are subjects of exploration and analysis. These mathematical branches provide further evidence of how infinity is an integral part of our mathematical framework.

In conclusion, it is crucial to revise our perception of infinity as a final destination or an unattainable value. Instead, we should recognize that the numbers we commonly use, including 1, 2, 3, 4, and even larger numbers like 100, are intimately linked to the concept of infinity. They emerge from infinite collections of 1's and remain embedded within the vast expanse of infinity. By embracing this understanding, we gain a deeper appreciation for the intricate interconnectedness of numbers and the boundless nature of mathematics. Infinity encompasses all numbers, both finite and infinite, and serves as a constant reminder of the unbounded possibilities and endless exploration that lie within the numerical realm.

Defining Infinity:

The definition of infinity is a concept that transcends the notion of a specific numerical value. Instead, infinity can be understood as the result of an infinite addition of the base number 1. To illustrate this, let us consider the values of numbers by repeatedly adding 1:

1 can be defined as the addition of 1, one time, resulting in 1
(1)

2 can be defined as the addition of 1, two times, resulting in 1 + 1   (2)

3 can be defined as the addition of 1, three times, resulting in 1 + 1 + 1   (3)

4 can be defined as the addition of 1, four times, resulting in 1 + 1 + 1 + 1   (4).

This pattern continues, where any given number n can be defined as the addition of 1, n times.

Now, let us explore the concept of infinity. Infinity (∞) can be defined as the addition of 1 an infinite number of times. It is represented as
1 + 1 + 1 + 1 + 1 + 1 + 1 + ...   (6)
Here, the ellipsis (...) represents an infinite continuation of adding 1.

This representation in equation (6) shows the natural form of infinity, where the sum of an infinite number of 1's leads to the concept of infinity itself.

To visualize this concept, we can consider a spherical representation of the number system. Imagine a sphere that encapsulates the entire number system. This sphere represents infinity, containing within it every number, sum of series, and mathematical concept.

Each number, whether it is 1, 2, 3, or any other value, exists within this infinite sphere. The sum of all numbers up to infinity is encompassed by infinity itself. Just as every number arises from the addition of 1, all numbers collectively form an infinite collection that resides within the concept of infinity.

By visualizing the number system in this way, we can comprehend the interconnectedness and inclusivity of infinity. It encapsulates every numerical value and serves as the vast expanse from which all numbers emerge and return.

Furthermore, this concept of infinity extends beyond the realm of whole numbers. It is also applicable to decimal numbers, fractions, and even irrational numbers. For instance, the decimal expansion of π (pi), a mathematical constant, goes on infinitely without repeating. This infinite decimal representation further emphasizes the presence of infinity within our numerical framework.

The spherical representation of the number system and the understanding of infinity as the sum of an infinite number of 1's allow us to grasp the profound nature of infinity. It is not a distant endpoint but a fundamental aspect of mathematics that permeates every numerical concept we encounter.

In conclusion, the definition of infinity lies in the concept of an infinite addition of the base number 1. Infinity encompasses all numbers and can be represented as the sum of an infinite number of 1's. By visualizing the number system as a sphere containing all numerical values, we gain a deeper understanding of the inclusive nature of infinity. It is not a finite destination but an infinite expanse from which numbers emerge and to which they return.

In conclusion, it is evident that any number present in the universe is encompassed within the concept of infinity (∞). We can generate arbitrarily large numbers by utilizing the infinite addition represented in equation (6). Similarly, any existing sum of series can be accommodated within the infinite expanse of ∞ by constructing larger and larger series.

Numerical Approach:

Calculating value of Infinity:

$$1 = \text{Addition of 1, one time} \quad = \quad 1 \tag{1}$$
$$2 = \text{Addition of 1, two times} = 1+1 \tag{2}$$
$$3 = \text{Addition of 1, three times} = 1+1+1 \tag{3}$$
$$4 = \text{Addition of 1, four times} = 1+1+1+1 \tag{4}$$

5 = Addition of 1, five times = 1+1+1+1+1   (5)

.

.

.

We can define Infinity, as
∞ = Addition of 1, Infinite times =
1+1+1+1+1+1+1+1+1+1+............   (6)
(6) represents the natural form of ∞.

By simplifying equation (6), we arrive at the expression:

$$\infty = 1 + 1 + 1 + 1 + 1 + 1 + 1 + 1 + 1 + \ldots$$

This implies that infinity (∞) is comprised of an infinite repetition of the base number 1. By expanding this expression, we can observe that ∞ can be represented as the sum of progressively increasing segments:

$$\infty = (1) + (1 + 1) + (1 + 1 + 1) + (1 + 1 + 1 + 1) + (1 + 1 + 1 + 1 + 1) + \ldots$$

This infinite summation highlights the nature of ∞ as an ever-expanding and all-encompassing entity.

To further illustrate the relationship between ∞ and numerical values, let us consider the expression (7):

$$\infty = 1 + 2 + 3 + 4 + 5 + \ldots$$

And from here we can see *the sum of all number upto infinity is nothing other than infinity*

Here, we have substituted the values of 1, 2, 3, 4, 5 from equations (1), (2), (3), (4), (5) into equation (6). This demonstrates that the concept of infinity (∞) is intimately linked to the progression of numbers. As we attempt to manipulate or alter the infinite expression (6) by multiplication, addition, subtraction, or division, ∞ retains its inherent value. It reaffirms the idea that infinity does not lie at the end of all numbers, but rather, all numbers emerge from ∞ and eventually return to it.

The conclusion we can draw from this exploration is that infinity (∞) is not a finite endpoint, but an infinite expanse from which numbers originate and to which they ultimately belong. Every number, every sum of series, and every mathematical concept is contained within the infinite framework of ∞. It is the fundamental essence that underlies the entire numerical system, guiding the way we perceive and manipulate numbers.

By understanding the true nature of infinity, we gain a profound appreciation for the boundless possibilities that mathematics offers. It opens up a world of exploration and discovery, where we can delve into infinite series, analyze infinite sets, and uncover the intricacies of calculus. Infinity serves as a reminder of the infinite potential and limitless horizons that exist within the realm of numbers.

In conclusion, infinity (∞) is not an unreachable destination, but rather a concept that encompasses all numbers and mathematical constructs. It is the origin and destination of every number, series, and mathematical concept we encounter. By embracing the concept of infinity, we unlock a deeper understanding of the interconnectedness and vastness of the numerical realm, leading to new insights and discoveries in the field of mathematics.

Mathematical Operations with Infinity:

Addition

We will do some calculations to see how numbers die or merge in infinity. We already do this calculation in our day-to-day mathematics problem solving and I hope you are also friendly with this calculation.

1. ∞+2 or (∞+n)

Let's explore some calculations to see how numbers merge or "die" in infinity. These calculations are familiar to us in our everyday problem-solving in mathematics. But now try to explore proof.

Consider the expression $\infty + 2$ or $(\infty + n)$, where n represents any specific number, such as 2. From equation (2), we know that 2 can be expressed as $(1 + 1)$. Additionally, equation (6) represents ∞ as an infinite repetition of 1.

Now, let's perform the calculation:

$$\infty + 2 = (1 + 1 + 1 + 1 + 1 + 1 + ...) + (1 + 1)$$

Expanding this expression, we can combine the terms:

$$\infty + 2 = \{(1 + 1) + (1 + 1) + 1 + 1 + 1 + 1 + ...\}$$

Simplifying further, we can group the terms:

$$\infty + 2 = \{2 + 2 + 1 + 1 + 1 + 1 + ...\}$$

Noting that the expression inside the curly brackets is similar to our previous representation of 2 as (1 + 1), we can substitute it:

$$\infty + 2 = \{(1 + 1) + (1 + 1) + 1 + 1 + ...\}$$

Continuing the simplification, we observe:

$$\infty + 2 = \{1 + 1 + 1 + 1 + 1 + 1 + ...\} \qquad (9)$$

Remarkably, we find that $\infty + 2$ is equivalent to ∞ itself. The sum of infinity (∞) and any specific number (in this case, 2) results in infinity. This demonstrates that when we merge any finite number with infinity, the outcome is still infinity.

This conclusion can be extended to other calculations involving infinity. For instance, if we multiply infinity by any finite number, the result remains infinite. Similarly, dividing infinity by a finite number or raising infinity to a finite power still yields infinity. In these operations, the finite numbers "merge" or "die" within the vastness of infinity, preserving its infinite nature.

These calculations reaffirm the concept that infinity is not a distant endpoint or a final destination for numbers. Instead, it is an all-encompassing entity that transcends finite values. Every finite number, no matter how large or small, becomes absorbed by infinity and loses its individual significance.

Infinity engulfs and absorbs finite numbers, incorporating them into its infinite expanse.

In conclusion, the calculations involving infinity reveal that when we combine or operate on infinity with any finite number, the result is still infinity. Numbers merge or "die" within the infinity, losing their individual identities. Infinity transcends the limitations of finite numbers and serves as the encompassing backdrop against which all numerical operations take place.

Understanding how numbers interact with infinity enhances our comprehension of the infinite nature of mathematics. It allows us to grasp the vastness and boundless potential inherent in the numerical realm. By recognizing the merging and transformative power of infinity, we can explore and appreciate the infinite possibilities that mathematics offers. From (2), we know that $2 = (1+1)$ and from (6), we know that $\infty = (1+1+1+1+1+1+ \ldots \ldots)$

Numerical Approach:

$$\infty + 2 = (1+1+1+1+1+1+ \ldots \ldots) + (1+1)$$
$$\infty + 2 = \{(1+1) + (1+1) + 1+1+1+1+ \ldots \ldots)$$
$$\infty + 2 = \{2+2+1+1+1+1 \ldots \}$$
$$\infty + 2 = \{(1+1) + (1+1) + 1+1+ \ldots \}$$
$$\infty + 2 = \{1+1+1+1+1+1+ \ldots \ldots\} \qquad (9)$$
$$\infty + 2 = \infty$$

2. $\infty + \infty$

In the case of $\infty + \infty$, we can refer back to equation (6), where ∞ is defined as the infinite repetition of 1:

$$\infty = (1+1+1+1+1+1+...)$$

Let's substitute this representation into the calculation:

$$\infty + \infty = (1+1+1+1+1+1+...) + (1+1+1+1+1+1+...)$$

Expanding the expression, we can group the terms:

$$\infty + \infty = \{(1+1) + (1+1) + (1+1) + (1+1) + (1+1) + (1+1) + ...\}$$

Simplifying further, we can combine the terms:

$$\infty + \infty = \{2 + 2 + 2 + 2 + 2 + 2 + ...\}$$

Noting that the expression inside the curly brackets is equivalent to the representation of 2 as (1+1), we can substitute it:

$$\infty + \infty = \{(1+1) + (1+1) + (1+1) + (1+1) + (1+1) + (1+1) + ...\}$$

Continuing the simplification, we observe:

$$\infty + \infty = (1+1+1+1+1+1+....) \qquad (10)$$

Surprisingly, we find that $\infty + \infty$ is equal to ∞ itself. The sum of two infinities (∞) results in infinity. This implies that infinity remains unchanged even when combined with itself.

This result demonstrates the nature of infinity as an unbounded concept that defies finite operations. Regardless of how many infinities are added together, the result remains infinite. Infinity encompasses and transcends finite quantities, rendering them insignificant in comparison.

An analogy that helps illustrate this concept is the idea of an infinite ocean. Imagine two infinite oceans merging together. No matter how vast each individual ocean is, their combination will still result in an infinite body of water. The infinity of the oceans swallows and incorporates any finite measurements within them.

Furthermore, this result can be extended to other operations involving infinity. Multiplying infinity by infinity ($\infty * \infty$), raising infinity to the power of infinity ($\infty^\wedge\infty$). The notion of infinity challenges our traditional understanding of arithmetic and reminds us of the limitless nature of mathematical concepts.

In conclusion, when we add two infinities ($\infty + \infty$), the result is still infinity (∞). This calculation exemplifies the expansive and unbounded nature of infinity. It emphasizes that infinity is not a final endpoint or a limit, but rather an encompassing concept that transcends finite quantities. The addition of infinities showcases the inherent inability of finite operations to alter the vastness of infinity.

Numerical Approach:

Again from (6), we know that $\infty = (1+1+1+1+1+1+......)$
Putting values,

$$\infty + \infty = \{(1+1+1+1+1+1+....) + (1+1+1+1+1+1+....)\}$$

$$\infty + \infty = \{(1+1) +(1+1) +(1+1) +(1+1) +(1+1) +(1+1) +....\}$$

$$\infty + \infty = \{2 + 2 + 2 + 2 + 2 + 2 +....\}$$

$$\infty + \infty = \{(1+1) +(1+1) +(1+1) +(1+1) +(1+1) +(1+1) +....\}$$

$$\infty + \infty = (1+1+1+1+1+1+....) \qquad (10)$$

$$\infty + \infty = \infty$$

Substraction

1. ∞ - 2

Let's explore the concept of subtraction involving infinity. We'll consider the calculation ∞ - 2 (or more generally, ∞ - n), using the representations from equation (6) and the value of 2 as (1+1).

∞ represents the infinite repetition of 1: ∞ = (1+1+1+1+1+1+...).

Substituting these values into the subtraction calculation:

∞ - 2 = (1+1+1+1+1+1+...) - (1+1)

Expanding the expression, we can group the terms:

∞ - 2 = {(1-1) + (1-1) + 1 + 1 + 1 + 1 + ...}

Simplifying further, we observe that (1-1) is equal to 0:

∞ - 2 = {(0) + (0) + 1 + 1 + 1 + 1 + ...}

Combining the terms, we get:

∞ - 2 = (1+1+1+1+1+...)

Notably, we have obtained the same representation of infinity as before. Therefore, ∞ - 2 is equal to ∞ itself.

This result demonstrates that subtracting a finite quantity from infinity does not diminish or alter the infinite value. Infinity remains unaffected by finite subtractions. The vastness of infinity encompasses any finite value and renders it insignificant in comparison.

You can easily understand this as if we send number 2 to make damage to infinity. As the number is 2 it can make only damage 2 to infinity but our infinity has power to generate as large as 1 so it will generate two 1's and get itself to its original shape of infinity.

Analogously, we can think of infinity as an endless universe, and subtracting a finite quantity from it is akin to removing a speck of dust. No matter how many specks are removed, the universe of infinity remains boundless and unchanged.

In conclusion, when subtracting a finite quantity from infinity ($\infty - 2$ or $\infty - n$), the result is still infinity (∞). This calculation emphasizes the boundless nature of infinity, which remains unaffected by finite subtractions. It underscores the idea that infinity is not a final destination or a limit, but an expansive concept that transcends finite quantities. Subtraction operations with infinity serve as a reminder of the incomprehensible vastness of this infinite concept.

Numerical Approach

$$\infty - 2 \text{ or } (\infty - n)$$

From (6) and (2) we know that $\infty = (1+1+1+1+1+1+....)$ and 2 $= (1+1)$

Putting values,

$\infty - 2 = (1+1+1+1+1+1+....) - (1+1)$

$\infty - 2 = \{(1-1) +(1-1) +1+1+1+1+....\}$

$\infty - 2 = \{(0) +(0) +1+1+1+1+....\}$

$\infty - 2 = (1+1+1+1+1+....)$

$\infty - 2 = \infty$

2. $2 - \infty$ or $(n - \infty)$

Let's delve into the concept of subtraction involving infinity in the context of $2 - \infty$ (or more generally, $n - \infty$), using the representations from equation (6) and the value of 2 as $(1+1)$.

Recalling that ∞ represents the infinite repetition of 1: $\infty = (1+1+1+1+1+1+...)$, we can now substitute these values into the subtraction calculation:

$2 - \infty = (1+1) - (1+1+1+1+1+1+...)$

Expanding the expression, we group the terms as follows:

$2 - \infty = \{(1-1) + (1-1) - 1 - 1 - 1 - 1 - 1 - 1 - ...\}$

Simplifying further, we observe that $(1-1)$ is equal to 0:

$2 - \infty = \{(0) + (0) - 1 - 1 - 1 - 1 - 1 - 1 - ...\}$

Combining the terms, we get:

$2 - \infty = (-1 - 1 - 1 - 1 - 1 - 1 - ...)$

At this point, we recognize the pattern of a repeated negative one (-1):

$2 - \infty = -1 \times (1+1+1+1+1+1+...)$

Now, using the representation of infinity (∞) as the sum of 1's, we can rewrite the expression:

$2 - \infty = -1 \times \infty$

Therefore, $2 - \infty$ simplifies to $-\infty$.

This result reveals an intriguing aspect of subtraction involving infinity. When subtracting infinity from a finite value, such as 2, the result is negative infinity. This signifies that the finite quantity is overwhelmed by the vastness of infinity, and its value is pushed to the negative infinite realm.

To illustrate this concept further, let's consider an analogy. Imagine having a jar with an infinite number of marbles (infinity). If we were to subtract two marbles from this infinite collection, the jar would still be filled with an infinite number of marbles. No matter how many marbles we remove, the jar remains infinitely full. However, the notion of a "negative infinity" arises as we attempt to quantify the difference between a finite value (2) and an unbounded quantity (infinity).

It is important to note that arithmetic with infinity can lead to diverse and sometimes paradoxical outcomes. In this case, the result of 2 - ∞ being -∞ illustrates the non-linear nature of infinity in mathematical operations. The notion of infinity challenges conventional arithmetic rules, emphasizing the need for careful consideration and understanding when working with infinite quantities.

In conclusion, subtracting infinity (∞) from a finite value (2) results in negative infinity (-∞). This demonstrates that finite quantities are overshadowed by the vastness of infinity, leading to a negative infinite value. Such calculations highlight the unique characteristics of infinity and its impact on arithmetic operations. Infinity defies simple subtraction and prompts us to explore the intricacies and limitations of numerical systems when faced with the concept of unboundedness.

Numerical Approach:

$2 - \infty$ or $(n - \infty)$

From (6) and (2) we know that $\infty = (1+1+1+1+1+1+....)$ and $2 = (1+1)$

Putting values,

$2 - \infty = (1+1) - (1+1+1+1+1+1+....)$

$2 - \infty = \{(1-1) + (1-1) - 1 - 1 - 1 - 1 - 1 - 1 - ....\}$

$$2 - \infty = \{0 + 0 - 1 - 1 - 1 - 1 - 1 - 1 - ....\}$$
$$2 - \infty = (-1 - 1 - 1 - 1 - 1 - 1 - ....)$$
$$2 - \infty = -1 \times (1 + 1 + 1 + 1 + 1 + 1 + ...)$$
$$2 - \infty = - \infty$$

3. $\infty - \infty$

To minimize confusion, let us assign the symbol "infinity1" to represent our natural infinity, from which we aim to subtract another infinity. Similarly, we will assign the symbol "infinity2" to represent the infinity we want to subtract from our natural infinity (now referred to as infinity1). It is crucial to note and remember this statement as it will be referenced in our subsequent discussions.

The Complexities of Subtracting Infinity: Unraveling the Ambiguity

Introduction:

Subtraction is a fundamental operation in mathematics, allowing us to find the difference between two numbers. However, when it comes to dealing with infinity, the concept becomes more intricate. Subtracting one infinity from another introduces ambiguity, resulting in two possible answers and an ongoing debate within the mathematical community. In this , we will explore the two potential outcomes of subtracting infinities and delve into the complexities that arise from this process.

Answer 1: ∞ - ∞ = 0 when both infinities are equal.

In certain scenarios, when infinity1 and infinity2 are considered equal, the subtraction yields a result of 0. This is rooted in the notion that if both infinities are identical, their infinite elements cancel each other out, leaving no remaining quantity. This conclusion may seem intuitive and straightforward, suggesting that the subtraction of equal infinities results in a null value.

Answer 2: The indeterminate form when infinity1 and infinity2 differ.

However, the situation becomes more intricate when infinity1 and infinity2 are unequal. In this case, the subtraction of infinities leads to an indeterminate form, where the outcome cannot be definitively determined. To understand this complexity, we need to consider the nature of infinity itself.

The Expanding Nature of Infinity:

Infinity possesses a remarkable characteristic: the power to continually expand and increase its magnitude. Its limitless nature defies conventional numerical rules and introduces challenges when attempting to subtract one infinity from another. Let us examine how infinity1 and infinity2 behave during the subtraction process.

Infinity1's Expanding Size:

When infinity1 is subtracted from infinity2, infinity1 seeks to preserve its infinity status by continually expanding its size. It counteracts the subtractive influence of infinity2 by generating an infinite number of 1's, effectively replenishing any reduction caused by the subtraction. This infinite growth process ensures that infinity1 remains infinitely large and maintains its distinct identity.

Infinity2's Equalizing Effort:

On the other hand, infinity2 endeavors to match the magnitude of infinity1. It strives to expand its own size to match that of infinity1, attempting to nullify the impact of subtraction. This equalizing effort perpetuates a never-ending cycle where infinity2 continually chases the expanding size of infinity1.

The Resulting Ambiguity:

Due to the dynamic nature of infinity and its capacity for limitless expansion, the subtraction of infinity1 and infinity2 fails to yield a definitive result. The continuous interplay between these infinities prevents the attainment of a conclusive answer. It is important to recognize that this ambiguity arises from the unique properties of infinity and does not adhere to the conventional rules of arithmetic.

Conclusion:

In the realm of mathematics, subtracting infinities introduces intricate challenges and two potential outcomes. When both infinities are equal, their subtraction yields a value of 0, aligning with our conventional understanding of subtraction. However, when the infinities differ, an indeterminate form arises, as infinity1 expands to maintain its infinite nature while infinity2 strives to equalize its magnitude. This perpetual back-and-forth creates an ongoing ambiguity, highlighting the complexities inherent in subtracting infinities. It serves as a reminder that the concept of infinity transcends our ordinary numerical operations and requires a nuanced understanding to navigate its intricacies.

Reality:

Let us consider a different perspective on this situation. What does infinity2 want? Referring back to our initial statement, we mentioned that we sent Infinity2 to cancel out infinity. Therefore, infinity2's sole purpose is to cancel out infinity, whether it is our original infinity (infinity1) or any other infinity. However, there is a significant change here. We know that our infinity1 has the power to generate an infinite number of ones, which implies that it can also generate other infinities by producing an infinite sequence of ones (as per the definition of infinity). Consequently, our infinity2 will cancel out the newly generated infinity, ensuring that the cancellation process takes place, and our infinity1 remains intact indefinitely.

The concept of developing infinity from infinity can be observed in various ancient scriptures and books that discuss the divine and supernatural realms. In these texts, gods are often portrayed as possessing the power to manifest in different forms or avatars. This ability to take on multiple appearances serves various purposes, such as confounding evil forces or demons (known as asuras). The parallel between this mystical phenomenon and the development of infinity from infinity can be intriguing to explore.

In many mythologies and religious traditions, gods are depicted as beings with immense power and transcendental attributes. They are believed to exist beyond the limitations of human perception and possess the ability to manipulate reality. One aspect of this divine power is the capacity to manifest in multiple forms or avatars. These avatars are distinct expressions of the same divine essence, each with its unique characteristics and purpose.

For example, in Hinduism, Lord Vishnu is regarded as the preserver and protector of the universe. He is said to have manifested in various avatars, such as Lord Rama, Lord Krishna, and Lord Narasimha, among others. Each of these avatars represents a different facet of Lord Vishnu's divine nature and serves a specific role in maintaining cosmic balance and defeating evil forces. By assuming these different forms, Lord Vishnu confuses and overpowers the asuras, who are constantly seeking to disrupt cosmic harmony.

Similarly, in Greek mythology, the god Zeus is known for his ability to transform into different entities or take on various appearances. Zeus would often adopt these disguises to interact with mortals or engage in battles with formidable opponents. This transformative power allowed Zeus to outwit his enemies and maintain his divine authority.

The parallel between these mythical accounts and the development of infinity from infinity lies in the idea of infinite possibilities and the transcendence of limitations. Just as gods are believed to possess the power to manifest in diverse forms, infinity is regarded as an infinite, boundless concept that defies conventional constraints. It is this notion of boundlessness that enables infinity to continuously generate new forms or expressions, much like the avatars of gods.

Moreover, these mythological examples also highlight the idea that the infinite nature of divinity cannot be diminished or exhausted. Despite assuming different forms or undergoing transformations, the essence of the divine remains eternal and infinite. Similarly, in the context of infinity, no matter how many infinities are generated or subtracted, the original concept of infinity remains unaltered and perpetually infinite.

While the comparison between the development of infinity from infinity and the divine avatars in mythology may be metaphorical, it allows us to contemplate the profound nature of infinity. It serves as a reminder of the vastness and incomprehensibility of infinity, as well as its capacity to continuously expand and transcend conventional boundaries.

In conclusion, the concept of developing infinity from infinity can be observed in various ancient scriptures and mythologies where gods are depicted as assuming multiple forms or avatars. These divine manifestations serve different purposes, including confusing and defeating evil forces. Although metaphorical, these accounts provide a fascinating parallel to the concept of infinity, emphasizing its boundless nature and its resilience in the face of attempts to subtract or diminish it. Exploring these parallels encourages us to contemplate the profound nature of infinity and its infinite possibilities.

Numerical Approach:

From (6) we know that $\infty = (1+1+1+1+1+1+....)$

$\quad\quad \infty = \{(1+1) +(1+1) +(1+1) +....\}$

$\quad\quad \infty = \{(2) +(2) +(2) +....\}$

$\quad\quad \infty = 2 \times \{1+1+1+1+1+1+....\}$

$\quad\quad \infty = \{(1+1+1+1+1+1+....) + (1+1+1+1+1+1+....)\}$

$\quad\quad (11)$

Putting values,

$\quad\quad \infty - \infty = \{(1+1+1+1+1+1+....) + (1+1+1+1+1+1+....)\} - (1+1+1+1+1+1+....)$

$\quad\quad \infty - \infty = (1+1+1+1+1+1+....) + \{(1+1+1+1+1+1+....) - (1+1+1+1+1+1+....)\}$

$\quad\quad \infty - \infty = \infty$

Multiplication

1. **2 × ∞ or (n × ∞)**

In the context of multiplication involving infinity, let's explore the expression 2 × ∞ (or more generally, n × ∞).

To begin, let's revisit the definition of infinity from (6) and the value of 2 from (2). We know that ∞ can be represented as an infinite sum of ones (1+1+1+1+1+1+...), and 2 can be represented as the sum of two ones (1+1).

Now, let's substitute these values into the expression:

2 × ∞ = (1+1) × (1+1+1+1+1+1+...)

Expanding the multiplication, we have:

2 × ∞ = (1 × (1+1+1+1+1+1+...)) + (1 × (1+1+1+1+1+1+...))

Simplifying further, we get:

2 × ∞ = (1+1+1+1+1+1+...) + (1+1+1+1+1+1+...)

This can be written as:

2 × ∞ = (1+1) + (1+1) + (1+1) + (1+1) + (1+1) + (1+1) + ...

Notice that we have a pattern emerging here. Each term in the sum represents the addition of two ones, which simplifies to 2. Therefore, we can rewrite the expression as:

$$2 \times \infty = 2 + 2 + 2 + 2 + 2 + 2 + \ldots$$

This can also be expressed as:

$$2 \times \infty = (1+1) + (1+1) + (1+1) + (1+1) + (1+1) + (1+1) + \ldots$$

Again, we see the same pattern as before. Each term in the sum represents the addition of two ones, resulting in 2. Hence, we can simplify the expression to:

$$2 \times \infty = 1+1+1+1+1+1+\ldots$$

This is precisely the definition of ∞. Therefore, we can conclude that:

$$2 \times \infty = \infty \qquad (12)$$

From this analysis, we see that when we multiply 2 by infinity, the result remains infinity. This can be generalized to any number (n) multiplied by infinity. The resulting product will always be infinity.

In conclusion, when multiplying a number (such as 2 or n) by infinity, the result is always infinity. This is due to the inherent nature of infinity and its ability to encompass and absorb any finite quantity. The concept of infinity introduces fascinating properties and challenges conventional arithmetic rules, reminding us of the vastness and infinite potential that this concept represents.

Numerical Approach:

$2 \times \infty$ or $(n \times \infty)$

Again from (6) and (2) we know that $\infty = (1+1+1+1+1+1+....)$ and $2 = (1+1)$

Putting values,

$2 \times \infty = (1+1) \times (1+1+1+1+1+1+....)$

$2 \times \infty = \{1 \times (1+1+1+1+1+1+....) + 1 \times (1+1+1+1+1+1+....)\}$

$2 \times \infty = \{(1+1+1+1+1+1+....) + (1+1+1+1+1+1+....)\}$

$2 \times \infty = \{(1+1) + (1+1) + (1+1) +(1+1) + (1+1) + (1+1) +....\}$

$2 \times \infty = \{(2) + (2) + (2) +(2) + (2) +(2) + ...\}$

$2 \times \infty = \{(1+1) + (1+1) + (1+1) + (1+1) + (1+1) + (1+1) +....\}$ $2 \times \infty = (1+1+1+1+1+1+....)$

$2 \times \infty = \infty$ (12)

2. $\infty \times \infty$

To explore the expression $\infty \times \infty$, let's recall the definition of infinity from (6), which states that ∞ is represented as an infinite sum of ones $(1+1+1+1+1+1+...)$.

Now, let's substitute this value into the expression:

$\infty \times \infty = (1+1+1+1+1+1+...) \times (1+1+1+1+1+1+...)$

Expanding the multiplication, we have:

$\infty \times \infty = \{(1 \times (1+1+1+1+1+1+...)) + (1 \times (1+1+1+1+1+1+...)) + (1 \times (1+1+1+1+1+1+...)) + ...\}$

Simplifying further, we get:

$\infty \times \infty = \{(1+1+1+1+1+1+...) + (1+1+1+1+1+1+...) + (1+1+1+1+1+1+...) + ...\}$

We can observe that each term in the sum represents the addition of an infinite number of ones. When we add infinite ones together, we still have an infinite number of ones. Thus, each term in the sum remains infinite. Therefore, we can rewrite the expression as:

$\infty \times \infty = \infty + \infty + \infty + ...$

Numerical Approach:

$\infty \times \infty$

From (6) we know that $\infty = (1+1+1+1+1+1+....)$
Putting values,

$\infty \times \infty = (1+1+1+1+1+1+1+....) \times (1+1+1+1+1+1+....)$

$\infty \times \infty = \{1 \times (1+1+1+1+1+1+....) + 1 \times (1+1+1+1+1+1+....) + ...\}$

$\infty \times \infty = \{1 + 1 + 1 + 1 + 1 + 1 +....\}$

$\infty \times \infty = \infty$

Division

1. ∞ ÷ 2 or (∞ ÷ n)

From previous discussion we know that ∞ = (1+1+1+1+......).
And now we want to divide infinity from number 2, you can take any number according to your will.

From (6) we know that ∞ = (1+1+1+1+1+1+....)

Putting values,

In this we will see the same nature as we have seen in subtraction. Because we know how subtraction and division are interrelated. To divide infinity with 2, we will write these in their basic form of existence which is in the form of 1's.

From (6) we know that ∞ = (1+1+1+1+1+1+....)

Putting values,

$$\infty \div 2 = \frac{(1+1+1+1+....)}{2}$$

$$\infty \div 2 = \frac{\{(1+1)+(1+1)+(1+1)+....\}}{2}$$

$$\infty \div 2 = \frac{(2+2+2+2+....)}{2}$$

$$\infty \div 2 = \frac{2 \times (1+1+1+1+.....)}{2}$$

$$\infty \div 2 = (1+1+1+1+1+1+....)$$

$$\infty \div 2 = \infty$$

From this we can observe that infinity is producing its 2 forms making it to double quantity so that dividing it does not affect infinity at all.

2. ∞ ÷ ∞

We will now try to divide infinity by infinity and this concept is same as subtracting infinity by infinity. Upper infinity will try to make its size large so that lower infinity will not be able to affect it on the other hand lower infinity will try to make its size equal to upper infinity to convert it to 1. let see the below solution;

From (6) we know that $\infty = (1+1+1+1+1+1+....)$

Putting Values;

$$\infty \div \infty = \frac{(1+1+1+1+....)}{(1+1+1+1+....)}$$

$$\infty \div \infty = \frac{\{(1+1+1+1+....)+(1+1+1+1+....)+(1+1+1+....)+....\}}{(1+1+1+1+....)}$$

$$\infty \div \infty = \frac{\infty \times \infty}{\infty}$$

$$\infty \div \infty = \infty$$

We can now see infinity is developing its infinite copies to get rid of other infinity who wants to convert it to one. After developing infinite copies upper infinity will become infinite times infinite which confuse lower infinite whom to cancel and upper infinity is remaining infinite.

Exponentials

The expression 1^{∞} has been a subject of debate and discussion among mathematicians, with different viewpoints on how to interpret its value. Some argue that it should be equal to 1 because multiplying 1 by itself any number of times will always result in 1. However, there is another perspective, which I align with, that suggests a different outcome.

This can be understood as multiplying 1 by an infinite times, effectively multiplying 1 infinite times:

1. 1^{∞}

$$1^{\infty} = (1 \times 1 \times 1 \times 1 \times 1 \times 1 \times \ldots)$$

To gain a better understanding, let's consider the nature of multiplication. Multiplication is often viewed as a shortcut for addition, where repeated addition can be represented more succinctly. In this case, we can express the same calculation as adding 1 a total of ∞ times, where ∞ represents infinity:

$$(1 \times 1 \times 1 \times 1 \times 1 \times 1 \times \ldots) = (1+1+1+1+1+1+\ldots)$$

This demonstrates that multiplying 1 infinite times is equivalent to adding 1 an infinite number of times, resulting in an infinite sum.

Using the equation (13) from the previous section, which shows that the infinite sum of ones is equal to ∞, we can conclude that:

$$1^{\infty} = (1+1+1+1+1+1+...)$$

And therefore:

$$1^{\infty} = \infty$$

In this context, the expression 1^{∞} evaluates to infinity, not 1. This interpretation aligns with the notion that the power of infinity is not constrained by the conventional rules of finite arithmetic.

It's worth noting that working with expressions involving infinity can lead to paradoxes and counterintuitive results. In mathematics, these situations often require careful analysis and consideration, and different viewpoints may exist. However, based on the reasoning presented here, the consensus favors the interpretation that 1^{∞} is equal to infinity.

Numerical Approach:

We know that $\infty = (1+1+1+1+1+1+....)$
Putting values,

$1^{\infty} = 1^{(1+1+1+1+1+1+....)}$

Means multiply 1 infinite times,

$1^{\infty} = (1 \times 1 \times 1 \times 1 \times 1 \times 1 \times ....)$

And we know that multiplication is the shortest form of addition, we can get same answer by adding 1, $(1 \times \infty)$ ∞ times.

So, $(1 \times 1 \times 1 \times 1 \times 1 \times 1 \times ....) = (1+1+1+1+1+1+....)$ $\qquad$ (13)

Using (13),

$1^{\infty} = (1+1+1+1+1+1+....)$

And hence,

$1^{\infty} = \infty$

2. 2^{∞} or (n^{∞})

2^{∞} means multiplying 2 infinite times.

From (6) we know that $\infty = (1+1+1+1+1+1+....)$

Putting values,

$2^{\infty} = (2 \times 2 \times 2 \times 2 \times 2 \times 2 \times ....)$

Taking 2 common,

$2^{\infty} = 2 (1 \times 1 \times 1 \times 1 \times 1 \times 1 \times ....)$

From (13), we know that $(1 \times 1 \times 1 \times 1 \times 1 \times 1 \times ....) = (1+1+1+1+1+1+....)$

Putting value,

$2^{\infty} = 2 (1+1+1+1+1+1+....)$

$2^{\infty} = (2 + 2 + 2 + 2 + 2 + 2 +.....)$

$2^{\infty} = \{(1+1) + (1+1) + (1+1) + (1+1)+....\}$

$2^{\infty} = \{1+1+1+1+1+1+....\}$

$2^{\infty} = \infty$

3. ∞^{∞}

From (6) we know that, $\infty = (1+1+1+1+1+1+....)$
Putting values,

$\infty^{\infty} = (1+1+1+1+1+1+....) \, (1+1+1+1+1+1+....)$

$\infty^{\infty} = \{(1+1+1+1+1+1+....) \times (1+1+1+1+1+1+....) \times$
$(1+1+1+1+1+1+....) \times .....\}$

And we know that multiplication is the shortest form of addition, we can get same answer by adding $(1+1+1+1+1+1+....)$ infinite times.

$\infty^{\infty} = \{(1+1+1+1+1+1+....) + (1+1+1+1+1+1+....) + ...\}$

$\infty^{\infty} = \{1+1+1+1+1+1+....\}$

$\infty^{\infty} = \infty$

Conclusion

1. From all discussion you can conclude that infinity is not a single number whose value can be determined; it is a special thing from which all the numbers, series and operations generate and die or merge. You can also see how infinity reacts when we approach it by adding or multiplying value to our number, it makes its size big and also how it reacts when we try to make its size small by subtracting and dividing it, it tries to expand itself. You can compare this infinity to the universe. How this universe behaves when we try to approach it or want to see it end, it becomes big as we can see it is expanding, the more the light reaches to the end of the universe the more objects we see.

Infinity is a concept that has fascinated philosophers, mathematicians, and scientists for centuries. It represents a limitless and boundless quantity that cannot be fully grasped or comprehended. While infinity is often associated with numbers and mathematical operations, it extends beyond mere numerical values and has profound implications in various fields of study.

When we consider infinity in the context of numbers, we realize that it is not a specific value that can be determined. Rather, it serves as a foundation from which all numbers, series, and mathematical operations emerge. Infinity is like the wellspring from which the entire numerical landscape originates and returns.

To understand how infinity behaves, let's examine its response to different mathematical operations. When we add or multiply a number by infinity, we witness its extraordinary capacity to magnify. No matter how large the initial number is, when we introduce infinity into the equation, the result becomes infinitely larger. For instance, if we multiply any finite number by infinity, the product will be infinite. Similarly, if we add any finite number to infinity, the sum will also be infinite.

For example, let's consider the equation $x * \infty = y$, where x is a finite number and ∞ represents infinity. As we increase the value of x, y grows without bound. Even if x is an unimaginably large number, the product with infinity surpasses it, demonstrating the incomprehensible nature of infinity.

On the other hand, when we attempt to diminish infinity's magnitude by subtracting or dividing, it resists our efforts and expands further. Infinity remains unaffected by these operations and maintains its boundless nature. Dividing infinity by any finite number still results in infinity, and subtracting a finite number from infinity remains infinite.

For example, let's consider the equation $\infty / x = y$, where x is a finite number. As we decrease the value of x, y approaches infinity without ever reaching a definitive value. This showcases the limitless nature of infinity and its refusal to be confined or diminished by conventional mathematical operations.

To illustrate the parallels between infinity and the universe, let's examine the behavior of the universe as we explore its vastness. When we observe the universe from our vantage point and attempt to comprehend its extent, we find that it expands as we venture deeper into its cosmic realms. The light that reaches us from distant galaxies reveals an ever-expanding universe, growing larger as we peer further into its depths.

Just as infinity cannot be contained within a finite number, the universe eludes our attempts to confine or comprehend its entirety. As we gain more knowledge and explore deeper into space, the universe's immensity becomes increasingly apparent, mirroring the boundlessness of infinity.

Moreover, the expansion of the universe provides another analogy to infinity. The concept of the Big Bang suggests that the universe originated from a singularity—an infinitely dense and small point—and has been expanding ever since. This expansion echoes the unbounded growth associated with infinity, further highlighting the profound connections between these two concepts.

2. One of the fascinating conclusions we can draw from the behavior of numbers and their relationship to the universe is the concept of unlocking possibilities. As we progress from one number to the next, we unlock new possibilities and expand our understanding of the numerical landscape. Similarly, when we approach the universe, it presents us with new dimensions and challenges us to explore further.

Let's consider the process of unlocking possibilities in numbers. When we approach the number 1, we gain the ability to move towards 2. This progression continues as we unlock subsequent numbers such as 3, 4, and so on. Each step forward unveils new numerical possibilities, expanding our comprehension of the number line.

For example, if we start with the number 1 and add 1 to it, we reach 2. By adding another 1, we can now reach 3. This sequential unlocking of numbers demonstrates the interconnected nature of numerical progression.

Interestingly, a similar pattern emerges when we examine the behavior of infinity. When we perform operations involving infinity, it generates as many "ones" as we require. This aligns with the unlocking of possibilities observed in the numerical sequence.

For instance, if we multiply a finite number by infinity, the result becomes infinite. This indicates that infinity has the power to generate an infinite number of ones. No matter how large the initial number is, infinity can magnify it indefinitely, akin to unlocking an infinite number of possibilities.

Drawing a parallel to the universe, we observe a similar unlocking process. As we approach the universe with any given quantity, it expands itself by that same amount, challenging us to reach further. This continuous expansion suggests that the universe is never-ending, as we are continually presented with new dimensions to explore.

Consider the concept of measuring the size of the universe. If we were to measure its extent and attempt to approach it, we find that the universe grows in proportion to our efforts. It invites us to delve deeper and discover more, much like the unlocking of possibilities in the numerical sequence.

However, there is an inherent limitation in our ability to approach infinity or fully comprehend the universe. Just as we cannot reach the infinite number line, we are unable to approach the universe in its entirety. Infinity represents a boundless concept that surpasses our finite understanding, and the universe, with its immense scale, presents a similar challenge.

If we accept the notion that the universe is infinite, then our inability to approach it becomes evident. We are limited by our finite nature, unable to grasp the entirety of an infinite entity. It becomes a perpetual quest, forever beckoning us to explore and expand our understanding.

In conclusion, the nature of numbers and their relationship to the universe reveals intriguing patterns and conclusions. The process of unlocking possibilities, observed in numerical progressions, mirrors the expansion and growth of the universe as we approach it. Just as approaching infinity generates an infinite number of ones, approaching the universe challenges us with new dimensions and endless possibilities. However, the inherent nature of infinity and the vastness of the universe prevent us from fully comprehending or approaching their entirety. The parallel between the behavior of numbers and the workings of the universe invites further exploration and contemplation, stimulating our curiosity and inspiring us to push the boundaries of our understanding.

End Note

I hold the belief that the Universe is fundamentally constructed through mathematics, and that mathematics serves as the sole means of communication with it. Furthermore, I believe that our ancient scriptures or ancient books contain a wealth of knowledge that surpasses our current understanding.

I want to finish my whole lecture with few lines from
Brihadaranyaka Upanishad

ॐ पूर्णमदः पूर्णमिदं पूर्णात्पुर्णमुदच्यते
पूर्णश्य पूर्णमादाय पूर्णमेवावशिष्यते ॥
ॐ शान्तिः शान्तिः शान्तिः ॥

Meaning:

**Aum! That is infinite, and this (universe) is infinite. The infinite proceeds from the infinite.
(Then) taking the infinitude of the infinite (universe), It remains as the infinite alone.**
Aum! Peace! Peace! Peace!

www.ingramcontent.com/pod-product-compliance
Lightning Source LLC
Chambersburg PA
CBHW070444220526
45466CB00004B/1769